YOUR KNOWLEDGE HAS VALUE

- We will publish your bachelor's and
  master's thesis, essays and papers

- Your own eBook and book -
  sold worldwide in all relevant shops

- Earn money with each sale

Upload your text at www.GRIN.com
and publish for free

Bibliographic information published by the German National Library:

The German National Library lists this publication in the National Bibliography; detailed bibliographic data are available on the Internet at http://dnb.dnb.de .

Imprint:

Copyright © 2019 GRIN Verlag
Print and binding: Books on Demand GmbH, Norderstedt Germany
ISBN: 9783668935037

This book at GRIN:

https://www.grin.com/document/465169

Lennart Weiss

The Influences of Lignite Mining on Grevenbroich, the Energy Capital of Germany

GRIN Verlag

Research Paper Geography

Lennart Weiss

Influence of lignite mining on Grevenbroich – the energy capital of Germany

---

Table of Content

1. INTRODUCTION

"Grevenbroich- the energy capital of Germany". This self-given title is both highfalutin and still rather empty. Being familiar with Grevenbroich, one can barely imagine this small city as the federal capital of anything. The only unique feature might be its close connection to lignite mining. But what exactly makes Grevenbroich so proud of it that the mayor decided to welcome all visitors with this phrase? What does tie soft coal and Grevenbroich so firmly together? And why did his successor choose to abolish this label again? Shall I count myself lucky to live here, or do I have to be afraid of terrible side effects?

As an inhabitant of Grevenbroich, I am therefore personally interested in the influence of lignite mining on my home town, which I am going to focus on in this research paper. For this purpose, the topic is essentially divided into two main parts.

At first, I am going to portray the social development having taken place consistent with the different phases of valorisation of Grevenbroich's mines, from the very first excavation to modern times. Naturally, the changing situation of Grevenbroich's population has entailed drastic demographic changes as well.

However, it is essential to take the huge ecological impact of brown coal extraction into consideration in order to gain a complete impression. Hereby, it has to be differentiated between the aquatic ecosystems and those ashore.

Weighing these social and ecological effects, one can finally draw a conclusion of the influence of lignite mining on Grevenbroich, the energy capital of Germany.

2. SOCIAL DEVELOPMENT - THE PHASES OF VALORISATION

In order to understand lignite mining's social influence on Grevenbroich, one necessarily has to consider the economic, demographic and urban development within the different phases of valorisation.

2.1 Small-scale mining

Given that Publius Cornelius Tacitus already reported "fire erupting from the earth"[1] in Cologne in 58 AD, the usage of lignite as fuel for building material seems to date back to ancient Roman times, which can also be proven by excavated ceramics from that era. Nevertheless, this process apparently, like many other antique achievements, fell into oblivion after the Roman Empire.

Only subsequent to the Thirty Years' War from 1618 to 1648, when the various efforts to remedy the multiple damages done by fighting rapidly raised the overall demand for wood as a raw material of vital necessity for construction, some parts of the poorer rural population resorted to the heating material turf from the marsh districts scattered all over Grevenbroich, but soon diminished past the reconstructions.[2]

By the time of the 18th century, dilletante forestry in forms of over-clearing had again more and more led to a fundamental lack of firewood which was why there only was an "annual per capita wood supply of 1.5 $[m^3]$"[2]. In contrast, the supply of peat was plentiful and "with $23m^3$ of turf a farmer could easily get along and could even sell some of it"[3]. Thus, the trove of lode within the scope of spring digging in Neurath resulted in the "first pre-industrial phase of lignite mining"[2].

Consequently, the surface coal deposits were soon exhausted. However, the miners, above all those in Grevenbroich and the surrounding area close to the border to the Netherlands, could certainly profit from the knowledge acquired by the Dutch who used to practice the so called "'Klütten' by accumulating earthy brown coal to a pulp, compacting it into conoid wooden forms to let it dehydrate"[4], for which process many inhabitants of the city could find the material needed on their own property. After all, the wealth of Grevenbroich's population

---

[1] Publius Cornelius Tacitus; annals XIII (58 AD)

[2] Alexander Weuthen – Die Geschichte des Braunkohletagebaus im rheinischen Braunkohlerevier und seine ökologischen und sozialen Auswirkungen (2015)

       2.1) Chapter 3: Der Braunkohleabbau im rheinischen Braunkohlerevier von den Anfängen bis 1945 (cf. pp. 14-15)

       2.2)/2.3) Chapter 3.1: Die Braunkohle im Zeitalter der Industrialisierung (p.19)

[3] Dau, J.H. Chr. – Neues Handbuch über den Torf (Leipzig, 1821)

[4] Dr. Johannes Demmer on behalf of Landesplanungsgemeinschaft Rheinland - Rheinisches Braunkohlengebiet: *Wirtschafts- und Sozialstruktur Teil 1* (1962)

       Part 2: Untersuchender Teil

       Chapter 1: Die wirtschaftliche Entwicklung unter dem Einfluß der Braunkohle

          Subitem 1.2: Die Entwicklung der Braunkohleindustrie

             1.21: Allgemeine Übersicht (cf. p. 12)

increased since the French mining law allowed the people to use the reserves on their land as concessions until 1865.[5]

Hence, the soft coal reserves kept the inhabitants of Grevenbroich from the timber shortage's most devasting repercussions.

2.2 Pre-Industrial Exploitation

Nevertheless, like in most other regions, Grevenbroich's community treasury was in 1700 still burdened with depts piled up during the fighting of 1618 to 1648 and the post war period. As a result, the local landed gentry intended to "rehabilitate the […] finances"[6] by commercialising peat decomposition in such a way as to use the profit for paying the creditors off. Therefore, Grevenbroich's government organized the sale of turf and likewise marketed the ash "declaring it an excellent fertilizer"[6].

Year	Earnings in "Reichstaler"	Comparable Purchasing Power (rounded)[i]
1711	361	330 000 €
1713	404	370 000 €
1714	431	394 000 €
1715	637	583 000 €

Document 1[i]

Having a closer look at document 1 which provides information on the city's earnings from the purchase, it becomes obvious that the peat mining's "dual use"[7] turned out to be a complete success enabling the city to get into the black again. Thence, Grevenbroich could benefit from this advantage over many competing regions.

2.3 Intensive mining in the course of Industrialisation and its aftermath

As for Grevenbroich's development, various aspects have since combined to lead to ongoing industrialisation.

In view of the mining's positive economic influx, the steadily increasing exploitation gradually let the workers come across more and more lignite reserves. Correspondingly, the number of jobs mounted from 7 permanent employees in the pits of Neurath in pre-industrial times to more than 13 000 workers in the Rhenish brown coal mines of the early 20th century.[5][8]

---

[5] Dr. Peter Zenker – Braunkohle, Kraftwerke, Briketts: *Der Norden des Rheinischen Braunkohlereviers*
  Part 1: Braunkohlenbergbau in Neurath; Chapter 2.1: Französisches Bergrecht (p.15)
  Chapter 8 : Erster Braunkohlenabbau in Neurath (p.34)
[6] Dr. Peter Zenker – Braunkohle, Kraftwerke, Briketts: *Der Norden des Rheinischen Braunkohlereviers*
  Part 4: Torf saniert die Gemeindefinanzen in Frimmersdorf/Neurath
  Chapter 6: Torf saniert die Gemeindefinanzen (p.258)
[7] J. Bremer: Das Kurkölnische Amt Liedberg (Mönchengladbach, 1930)
[8] Dr. Johannes Demmer on behalf of Landesplanungsgemeinschaft Rheinland - Rheinisches Braunkohlengebiet: *Wirtschafts- und Sozialstruktur Teil 1* (1962) (p.15)

On account of the higher amount of people involved, more and more mining law trade unions, which, as a corporate business form, are reserved to the mining sector, erupted in line with the further allotments.[9] As the employee representation bodies also operated as joint stock companies following the interest of financial gain, they were additionally eager to attract workers by means of wages above average as well as several social perks. [10]

For instance, the trade unions shaped the Prussian mining law in such a way as to improve both, the working conditions as well as the harmonic and just cooperation with the population with regard to concessions and mining property.[9] Moreover, the introduction of the eight working hours a day that are nowadays regarded as common can be put down to the Stinnes-Legien-Agreement enforced these labour unions. [11]

As for the aim of ameliorating the labourers' circumstances, it gave another rise to workforce, likewise permitting expansion in mining activity. Thus, these positive side effects mutually entailed enduring mechanisation, which enhanced the workers' duties along with the opportunities to extract lignite from deeper layers of earth and to compact it to coal briquets. [10]

Furthermore, one must not neglect the importance of electricity generation by means of soft coal.

All these developments have, little by little, paved the way for supplementary industrialisation in terms of sequential industries such as the exploitation of diverse raw materials from the soil of the open pit mine. In addition to that, the conversion of lignite into electricity has since allured many energy-intensive productions like the fabrication of sugar from the beet roots the region is known for and the chemical sector. Besides, metal processing and the accretive technical automation have supportively accompanied each other. Whilst lignite mining provided the metal engineering industry with sufficient power supply, the latter accommodates the strip mines with equipment. Henceforth, "since capital goods in particular are delivered to the new lignite mines and the energy production based on it from machinery factories in the wider area, indeed from all over Germany, soft coal mining also promotes the German machinery industry to a considerable extent" [12].

---

[9] Dr. Peter Zenker – Braunkohle, Kraftwerke, Briketts: *Der Norden des Rheinischen Braunkohlereviers*
    Part 1: Braunkohlenbergbau in Neurath; Chapter 2.3: Bergrechtliche Gesellschaften (p.16)
        Chapter 2.2: Preußisches Bergrecht (p.16)
[10] Dr. Johannes Demmer on behalf of Landesplanungsgemeinschaft Rheinland - Rheinisches Braunkohlengebiet: *Wirtschafts- und Sozialstruktur Teil 1* (1962)
    Chapter 1.23: Die Entwicklung der Produktion , der Beschäftigten, der Umsätze, der   Arbeitszeiten und Arbeitsentgelte (p.16)
    Chapter 1.2(1): Die Entwicklung der Braunkohleindustrie - Allgemeine Übersicht
[11] Alexander Weuthen – Die Geschichte des Braunkohleabbaus im rheinischen Braunkohlerevier und seine ökologischen und sozialen Auswirkungen
    Chapter 3.3: Die Rolle der Braunkohle zwischen 1913 und 1933 (p.30)
[12] Deutscher Braunkohlen-Industrie-Verein E.V. 1885.1960
    Part 4: Braunkohle und Öffentlichkeit
        Chapter 1: Die Braunkohle und die Gemeinden (pp. 74-75)

Although the large land area occupied by the pits suggests a negative development in the agricultural sector, the direct opposite is in fact the case. Though the number of farmers in the once agrarian region declined to less than 4% of the labouring inhabitants, the erstwhile cultivators did not become unemployed, but usually sought and found well paid jobs in industrial segments. Still, the remaining agricultural holdings addressed "by far more than 80 000 consumers"[13] in Grevenbroich's consistently more urbanised trading area. After all, mining the area meant a valorisation to an earning-capacity value of 3000% of agricultural use. "So, [...] lignite mining constitute[d] the direct and indirect agricultural spine of these zones"[13].

Branch of Industry	Staff
brown-coal extraction and agglomeration	22,500
power generation on the basis of lignite	3,500
glass- and chemical industry on the basis of lignite	10,000
electrometallurgy on the basis of lignite	2,500
electrochemistry	7,500
TOTAL	46,000

Document 2   Employees in the Rhenish soft coal exploitation and its sequential industries[ii]

The table in document 2 clearly shows the essential significance of soft-coal related industries, providing 46,000 jobs in the northern subarea with a total population of a rounded up 74,000 people.[14] Interestingly, the associated industries only deceeded the actual lignite mining and conversion by 6,000 employees, underlining the diversification of the broadly-based branches of industry, already in the later phase of industrialisation, by which time only 36.2% of the industrial workforce was not connected to lignite.[14] Respectively, the loss of work places in the wake of incremental mechanisation was most certainly outnumbered by the creation of jobs in newly settled businesses. Strictly speaking, these sections often offered even better payment. To give an example, the iron and steel industry then paid an hourly wage of about 300 pennies in comparison to an average of 266 in the whole German industry.[14] It follows logically that soft coal extraction has left a primary mark on Grevenbroich's employment structure. At least, 1665 people out of the 2007 inhabitants of Grevenbroich's district Neurath were employed in industries connected to lignite, ensuing high trade tax revenues.[14] To that effect, the community could also partake in the people's increasing wealth which was why "the tax receipts of the industrial municipalit[y] allowed the construction of generous public

---

[13] Deutscher Braunkohlen-Industrie-Verein E.V. 1885.1960
    Part 4: Braunkohle und Öffentlichkeit
        Chapter 1: Die Braunkohle und die Gemeinden (p. 75)
[14] Dr. Johannes Demmer on behalf of Landesplanungsgemeinschaft Rheinland - Rheinisches Braunkohlengebiet: *Wirtschafts- und Sozialstruktur Teil 1* (1962)
        Chapter 1.3(1): Die Folgeindustrien/ Allgemeine Übersicht (p.18)
        Chapter 1.65: Berufszugehörige und Erwerbspersonen in der Land- und Forstwirtschaft (p.33)
        Chapter 1.4: Die übrige Industrie (Nichtfolgeindustrie) (p.23)
        Chapter 1.23: Die Entwicklung der Produktion, der Beschäftigten, der Umsätze, der Arbeitszeiten und Arbeitsentgelte (p.16)
        Chapter 3.51: Verkehrspflege und Verkehrsförderung [...] (p.83)
        Chapter 2.6: Das Teilgebiet Nord – *Neurath/Frimmersdorf/Grevenbroich/Gustdorf (pp. 49-50)*

buildings, modern schools, churches, sports grounds and gardens, swimming baths and the maintenance of commendable path networks and sewage systems"[15], enhancing the population's standard of living. The "Erasmus-Gymnasium"[16] and the foundations for the remodelling of the city's parks that later accommodated the state horticultural show can be ascribed to the flourishing economic situation of that time.

As a matter of course, the soft-coal, the briquets and all the products of the multiple other industries had to be exported. Consequently, the enterprises invested into different means of transport, to give an example, rail links like the north-south railway[17] or "a dense and efficient road network"[15]. Anyway, "through the technical progress and the invention of the railway the demand for heating material amplified as much as the supply" [18]. Therefore, this virtuous circle fundamentally amended Grevenbroich's infrastructure, attracting even more enterprises. Moreover, the revised transport connection simplified the daily in-commuter flow of more than 3,000 people from the environs. In addition, the "in all senses very good" [15]traffic situation facilitates the "more distinct reciprocity in the exchange of workforces [due to the] versatil[ity] of [Grevenbroich as a "highly industrial site"] and the industry of neighbouring locations"[15], promoting further progress in the agglomeration of industries in Grevenbroich as a key region.

2.4 Opencast mining in post-industrial times

Whereas most other industries were harshly damaged by the Franco-Prussian-War and the two world wars, the state thrived to protect lignite mining as Germany's main way of power supply. Hereinafter, forced syndicates were to ensure the gain of electricity, preventing about 30% of the miners from conscription calls. However, the impact of the enabling act led to a temporary neglect of financial gain and the same applied for the French occupants. [19] At least, the surface mines in Grevenbroich and its surroundings were, unlike in many other areas, largely spared by the Allied forces because of its proximity to France and the resultant possible use for the country.[20] Therefore Grevenbroich was less scratched by air attacks, so that the soft coal mining could soon recover for which reason "the volume of coal production before

---

[15] Dr. Johannes Demmer on behalf of Landesplanungsgemeinschaft Rheinland - Rheinisches Braunkohlengebiet: *Wirtschafts- und Sozialstruktur Teil 1* (1962)
        Chapter 2.7: Übersicht über das Teilgebiet Nord (p.56)
        Part 3: Anhang-Gemeindestatistik
            Section 1.3: Teilgebiet Nord; Grevenbroich (p.109)
        Chapter 2.6: Das Teilgebiet Nord – *Neurath/Frimmersdorf/Grevenbroich/Gustdorf (pp. 49-50)*
[16] Erasmus Gymnasium Grevenbroich: Schulgeschichte
[17] Westdeutsche Wirtschaftsmonographie Folge 2: Braunkohle (p.47)
        Bundesbahn-Oberrat Kurt Rauch - Die Nord-Süd-Bahn im Rheinischen Braunkohlerevier
[18] Alexander Weuthen – Die Geschichte des Braunkohleabbaus im rheinischen Braunkohlerevier und seine ökologischen und sozialen Auswirkungen
        Chapter 3.1: Die Braunkohle im Zeitalter der Industrialisierung (p.20)
[19] Alexander Weuthen – Die Geschichte des Braunkohleabbaus im rheinischen Braunkohlerevier und seine ökologischen und sozialen Auswirkungen
        Chapter 3.3: Die Rolle der Braunkohle zwischen 1913 und 1933 (p.26)

[20] Stadt Grevenbroich

the First World War was already reached again in 1918"[21]. One of the purposes for this lay in the fact that "extensive reparations in stone coal"[21] necessitated a higher supply of lignite as the only sufficient domestic energy source. Against this context, technical inventions such as the introduction of overburden excavators and bucket wheels made the open cast mines in Grevenbroich more competitive, encouraging the city to proclaim itself the "energy capital of Germany".

Notwithstanding, war damages and cumulative urbanisation involved relatively high rates of housing shortage, especially among the newly arrived workers. In order to elevate the living standard and the attending satisfaction and employee morale, the state and the companies "support[ed] and propagate[ed]" the creation of about 9,000 homesteads in so called small holdings for the pitmen in the Rhenish brown coal fields, some of which still mark some parts of Grevenbroich like Neuenhausen.[22] Furthermore, the firms offered additional social contributions like "overtime pay, family income supplement,[…] vacation compensation, [insurances for] accidents or sickness, wage adjustment for holidays during the week, patronage rewards, Christmas bonusses and allowances on the occasion of births, communion and confirmation"[23] as well as continuative education as a basis for graduation from university pursuing the option of promotion.[22]

Whilst some could benefit from the edification of social housings, others lost their homes on account of relocation. So far, the roughly 33,000 residents of about 50 villages have been confronted with the issue of their resettlement yet. Nowadays, only the name of the well-known cavity still reminds of the former parish of Garzweiler. There is no doubt that your home being razed to the ground is a thing that is really difficult to cope with, above all for older generations losing their footing, which can barely be compensated by adjustment payments. [24]

What might count among the most major implications is the destructions of cultural heritage through the relocation of mediaeval villages like Reisdorf, where the ancient cloister "Sankt Leonard" had been open to Reisdorf's believers for more than 500 years.[25]

---

[21] Alexander Weuthen – Die Geschichte des Braunkohleabbaus im rheinischen Braunkohlerevier und seine ökologischen und sozialen Auswirkungen
      Chapter 3.3: Die Rolle der Braunkohle zwischen 1913 und 1933 (p.25)/ (pp.26-27)
[22] Deutscher Braunkohlen-Industrie-Verein E.V. 1885.1960
      Part 4: Braunkohle und Öffentlichkeit
         Chapter 4: Das Eigenheim für den Bergmann
      Part 2: Mensch und Arbeit im Braunkohlenbergbau
         Chapter 2: Nachwuchsplanung in den einzelnen Braunkohlen-Revieren
[23] Westdeutsche Wirtschaftsmonografie Folge 2: Braunkohle
      Dr. Hellmut Bauer – Betriebliche Sozialleistungen im rheinischen Braunkohlerevier (pp. 75-76)
[24] C. Lötgers and E. Mayers-Beeks – The social compatibility of Resettlement Schemes in the Rhenish Lignite Mining Area in a setting of Social Change (p. 363)
[25] Dr. Peter Zenker – Braunkohle, Kraftwerke, Briketts: *Der Norden des Rheinischen Braunkohlereviers*
      Part 2: Braunkohlenbergbau in Frimmersdorf
         Chapter 13: Umsiedlung und Erftverlegung (p.187)
         Chapter 14: Archäologische Funde (p.196)

Nonetheless, strip mining has also unearthed archaeological discoveries equally charged with history. For example, there was "the 'Husterknupp' [from 1080] in Frimmersdorf, the Frankish grave in Morken [and] the equipment of a shaman in Königshoven"[25].

To come to an interim conclusion, the positive side effects of lignite mining's social influx on Grevenbroich definitely prevail despite some negative consequences, since open pit extraction and its subsequent industries still dominate the city's economy, which becomes apparent referring to RWE as a major employer of the region.

2.5 Demographic Development

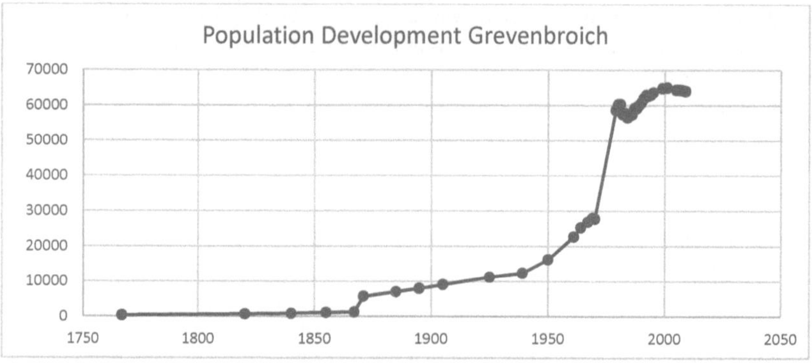

Document 3 Population Development of Grevenbroich[iii]

The cited graph in document 3 shows Grevenbroich's population development from 1767 to 2009.

Between the second half of the 18th century and 1867 only a very slight increase of less than 1 000 inhabitants took place. When the industrial revolution started the population experienced a sudden rise of 4 500 people within a time span of four years because Grevenbroich gained more and more importance as an industrial centre. As depicted above, the change Grevenbroich had undergone did not only attract potential employees but also those seeking to profit from the advantages the city could now shine with. This was the reason for the steady populational increase from 5 750 people in 1871 to 12 380 persons in 1939, many of whom from foreign countries, which can be put down to the establishment of further mines and power stations. Given that lignite had a significant share in the city's public welfare, the amount of inhabitants ascended even more steeply until the former administrative district had reached a population of nearly 28 000 in 1970, by which time Grevenbroich had become such an important industrial location that it then gained another 30 000 persons through incorporation of smaller surrounding municipalities until 1979. Since then, Grevenbroich's population has reached its peak of more than 65 000 people in 2001 and has been subject to a rather fluctuating performance typical of industrialised regions and their demographic change.

All in all, it is evident that lignite mining has tremendously effected Grevenbroich's social structure making it an industrial base of great importance for the whole Rhenish brown coal field.

3.ECOLOGICAL IMPACT

Although the economic and demographic influences might be more prominent as they have most directly affected Grevenbroich's population, it goes without saying that lignite mining has also had a huge impact on the environment in and around the city. "[A]s an open pit mine [, the operations were] coerced to absorb the [...] surface of the earth, hills and valleys, forests and fields [...] and actually everything"[26]. Manifold counter measures have already been taken. Though, the outcome of recultivation has to be assessed in each individual case.

3.1 Landscape

"[The] inevitable and most conspicuous effect of mining is the destruction of the landscape in the immediate mining field"[27]. Next to the various power stations, repositories or high-voltage transmission lines forming the landscape, RWE alone has dug more than 201 km^2 in the Rhenish lignite district, most of which comprised agricultural, forestal and water areas.[28] After all, lignite mining therefore necessarily affects all the surrounding territory.[26]

Consequently, the whole land area has changed due to the numerous interventions.[5.2]

In addition to that, repeated relocations of the river Erft have also left their mark in the so called Erfttal.[5.3]

3.2 The Land Area's Ecosystem

As for the land area, excavation material has shaped the landscape in terms of spoil heaps. By means of example, the 120 m above the terrain of the Vollrather Höhe make the mine dump of 465 000 000 m^3 the highest hill in Grevenbroich, by far exceeding the height of the Welchenberg as the city's largest natural elevation. [29]

Besides, nearly 90 km^2 of waterbodies and woods have been destructed within the framework of large-scale mining.[30] Naturally, this meant a drastic reduction of habitats, especially for typical woodland species such as boars or roe deer and waterfowl.

Nonetheless, nature conversation and the protection of species which are often erroneously equated must be distinctly differentiated. According to Professor Werner Kunz, populist

---

[26] Westdeutsche Wirtschaftsmonografie Folge 2: Braunkohle
       Prof. Dr. Theodor Kraus – Der Braunkohlenbergbau und die niederrheinische Landschaft (p. 16)/ (p. 18)
[27] W. Konold, R. Böcker, U. Hampicke – Handbuch Naturschutz und Landschaftspflege
       Chapter 7: Land- und Wassernutzung und Naturschutz
           Subtopic 10.1: (Ulf Dworschak, Udo Rose-) Das Rheinische Braunkohlerevier (p.5)
[28] RWE Power im Rheinischen Braunkohlerevier (p.6)
[29] W. Konold, R. Böcker, U. Hampicke – Handbuch Naturschutz und Landschaftspflege
       Chapter 7: Land- und Wassernutzung und Naturschutz
           Subtopic 10.1: (Ulf Dworschak, Udo Rose-) Das Rheinische Braunkohlerevier (pp.5/15)
[30] RWE Power im Rheinischen Braunkohlerevier (p.6)

conservationists and lurid media convey a misleading impression of the situation, leading to general demand for misdirecting actions. He is of the opinion that the wrong assumption of nature's and assortment's preservation being synonymous majorly result from a deficient understanding of the glacial period's after-effects. For the expansion of the drift epoch's core areas in Asia Minor's velds, the southern Balkans and the Mediterranean regions has directly struck central Europe, the former vegetation of that time has persistently been replaced by light forests, many local animals and plants have since depended on continuous natural destruction in form of hurricanes, floods or conflagration. Gradually, the flora and fauna of Grevenbroich have also adapted to the biotope's transformation at regular intervals, above all since the Stone Age when the people living here successfully started to subdue their environment for their own purposes. Thus, the perception of "untouched nature" and primeval forest is, at least in Grevenbroich, sheer imagination. [31]

In fact, Professor Werner Kunz is convinced that lignite mining's impact on Grevenbroich's ecosystem should be rated "rather positively"[31]. Owing to the agricultural use of more than half of the removed area[30], the intense use of fertilisers like ammonium has led to an extremely high nitrogen content in Grevenbroich's soils. Certainly, only 10% of the local plant types are nitrophilous and therefore supplanted other species. Through denudation and shifting of the layers of earth, loess loam was unearthed that was "still charged with alkaline cations and feature[d] all positive characteristics of [...] topsoil"[32], including advantageous microbiological fauna upgrading the soil's quality.[29] With an average layer of 120 cm of loess, the maximum plate number 80 of the high lime soil covering "exiguous sand and gravel surfaces"[33] grants preeminent capability of aggregating water and nutrients likewise, making the soil extremely fertile.[34] In this vein, chalky soil sward, which usually only occurs in the Alpine scenery[35], supplied a basis for 16 rare kinds of orchids as well as pines. As the previous disposal of plants caused very little competition, these preconditions let such species gain ground.

The new vegetation, suddenly dominated by sun-heated field, provided new habitats for various kinds of reptiles, additionally attracting raptors and other birds such as larks, pipits and shrikes, which are all itemised on the Red List.[36]

Besides, acclivities like the aforesaid Vollrather Höhe created escarpments that would be rather uncommon in the Erft's meadows. On these ideal biotopes for insects preferring locations on a slope, a multitude of bugs, above all spare butterflies and wasps, have settled, again serving as nourishment for other types. In the era of drastic bee mortality, it is also worth

---

[31] Biology Professor Werner Kunz
[32] Deutscher Braunkohlen-Industrie-Verein E.V. 1885.1960
       Part 4: Braunkohle und Öffentlichkeit
              Chapter 5: Die Wiederherstellung landwirtschaftlicher Böden (p.94)
[33] Westdeutsche Wirtschaftsmonografie Folge 2: Braunkohle
       Prof. Dr. Theodor Kraus – Der Braunkohlenbergbau und die niederrheinische Landschaft (p. 25)
[34] Deutscher Braunkohlen-Industrie-Verein E.V. 1885.1960
       Part 4: Braunkohle und Öffentlichkeit
              Chapter 5: Die Wiederherstellung landwirtschaftlicher Böden (p.94)
[35] Pott (Pflanzengesellschaften) – Syntaxonomie 1995
[36] Biology Professor Werner Kunz

mentioning that Grevenbroich's landscape hence poses a habitat perfectly suitable for these indispensable pollinators contributing to the whole area's fruitfulness. Recapitulatory, lignite mining has even enriched Grevenbroich's biodiversity.[36]

3.2.1 Recultivation of the Land Area

Regardless, the sudden emergence of environmental awareness because of climate change aroused interest in extensive recultivation among Grevenbroich's citizens in the late 1940s. Ergo, recultivation has since been pursued in many respects.

Exemplarily, Doctor Gero Vinzelberg, open pit planner of RWE Power, states grit as a remedy to fight abandoned vegetation before resettling animals, particularly rodents like hazel dormice or hamsters.[37]

By the same token, RWE has established a recultivation innovation station at Schloss Paffendorf in order to "restore the former area ratio"[38].

	Former Area	Recultivated Area
Agriculture	106.6 km^2	108.8 km^2
Forests	76.9 km^2	79.7 km^2
Water Bodies	6.4 km^2	8.0 km^2
Other	10.9 km^2	11.6 km^2
Sum	200.7 km^2	208.1 km^2

Document 4 – Recultivation of/around RWE Power's share of mining land[iv]

As depicted in the table of document 4, those measures have even increased the overall land area not lying fallow in the anterior mines.

Having a closer look, it is clearly to be seen that until know, forestry is the only section having lost some area. Against the context of restoration, this current situation shall be altered by means of large-area reforestation.[39]

Indeed, forest stand plays an important role in absorbing some of the carbon-dioxide emissions the power stations are often criticised for. Nevertheless, the biologist Professor Werner Kunz puts this approach down to populism, for "in general, people think the more trees are grown, the healthier the ecosystem."[40] In truth, this oftentimes turns out to be a fallacy. As described above, Grevenbroich's species favour a more diverse landscape. For instance, trees are grown to stop erosion which, in fact, is necessary for the insects' optimum on precipices. In addition to that, the monoculture of poplars "in row cultivations clearly stands out even visible to lays"[41] and is extremely susceptible to diseases which can easily spread and devastate whole tree populations.

---

[37] Doctor Gero Vinzelberg
[38] Doctor Gero Vinzelberg on behalf of RWE Power in the NGZ
       Tim Kronner - Restloch in Grevenbroich: Tagebau wird zur Naturlandschaft
[39] Doctor Gero Vinzelberg
[40] Professor Werner Kunz
[41] Westdeutsche Wirtschaftsmonografie Folge 2: Braunkohle
       Landforstmeister Wemper – Die Rekultivierung im Kölner Braunkohlenrevier (pp. 50/51)

To counter this obnoxious development, "herbaceous nitrogen collectors have many times been cultivated between the silviculture to improve the nitrogen supply"[41]. Although the nutrients might have strengthened this particular kind of tree, some local plants have again been dispersed in line with overgrowing.

The same effect is to be observed on the increased agricultural area of nearly 109 km^2 where consistent soil degradation is to be expected. Therefore, the population of resettled bats and house martins has once again diminished.[40]

Yet, aquatic birds can definitely benefit from the higher amount of water bodies of now 8 km^2.

All in all, the entire recovery of the original situation seems barely desirable, as it would lessen the richness in species.

3.3 Water Management and its influence

Speaking of it, the water management and its influence connected to lignite mining are not to be underestimated. Given that the strip mines go down to more than 200 m below the surface, it was inevitable to pump the ground water around the pits down.[42]

Correspondingly, "in several regions of [Grevenbroich and its vicinity], a groundwater draw-down of 300 m and more are no longer extraordinary"[43]. The consequential depressions sometimes lead to "tectonic faults"[43].

Still, there are most direct influences on the aquatic ecosystem. With respect to the Erft's meadows' wetlands, "the grip water saved by in the soil alone does not suffice anymore for an optimum feed of the plants"[43]. What is more, the groundwater recession drastically aggravated the conditions of typical floodplain species so that a "dieback of 90% of all alders [took place] because the water balance deficit could no longer be covered from ground water resources"[43], leading to the domination of new plants like the giant cow parsnip. Thence, local fauna also lacked in suitable habitats.

The amount of water immediately drained from the open pit mines is scilicet extended by heat exchange water needed in the coal-fired power plants, which is "at latest since 1965, to a preponderant portion ingested by the Erft"[43]. The river's mean natural flow of about 3 m^3/sec is enormously boosted by another 6 m^3 of drainage water/sec. At first, the conflux of clean groundwater had the positive effect of attenuating the sewerage polluted river. However, the melioration of the Erft, which already began 200 years ago, was intensified, transforming the river into a kind of canal in such a way as to prevent floods as a result of the higher water-level. Withal, the cooling water, which is heated up to more than 24°C during the heat transfer

---

[42] Doctor Udo Rose

[43] W. Konold, R. Böcker, U. Hampicke – Handbuch Naturschutz und Landschaftspflege
    Chapter 7: Land- und Wassernutzung und Naturschutz
        Subtopic 10.1: (Ulf Dworschak, Udo Rose-) Das Rheinische Braunkohlerevier (pp.6/7)

in the power station, raises the Erft's normal temperature of far below 10°C up to 13-15°C in winter. As a consequence, the unusual thermal determining factors gives an advantage to non-resident species.[44]

Although the competition might not have suppressed local flora and fauna yet, half of the 16 types of water plants in the Erft come from the subtropics, six of which depend on warm water temperatures. Plants from aquariums or ponds, such as the shellflower, have settled at the broaden banks, shading and likewise dislodging regional plants. The large quantities of the spiral wrack and the South Asian waterweed occur area-wide. In this competition, resident plants' acclimatisation to the prior situation disadvantages their populations. As for invertebrates, 10% of the species are non-territorial and often come from the Rhine. Specifically, the number of worms and leeches has retrograded.[44]

The modified biotopes have also rendered possible the settling of diverse fauna. The number of guppies living in a river part is positively dependent on the distance to the water induction points. More prominently, even piranhas have been detected. However, their existence has to be relativised in so far that both species only are exposed aquarium fish. Whilst the guppies' population is restricted to the Erft's warmest parts, the piranhas are very rare single cases, since they cannot breed without the optimum of 28°C.[44]

Howbeit, the overall water quality, especially the one of the groundwater, has downgraded, which "manifests [...] in the deviation of basins and the hardening of the water"[45]. Furthermore, excavation material has polluted the drainage water with harmful masses of iron and manganese.[46]

3.4 Recultivation of the Aquatic Ecosystem

Organised by the Erftverband, the cities, communes, counties and the RWE collectively finance different programmes of renaturation. The nature-orientated removal of the Erft is supposed to more diverse structures regarding habitats by renaturating the river's original course. Furthermore, the "Ministerium für Umwelt, Raumordnung und Landwirtschaft" (MURL) intends to provide 20 000 000 m³/a of substitute water in order to revive the Erft wetlands.[46] Doctor Udo Rose, head of the Erftverband, acts from the assumption that the competitive advantage of the invasive plants and animals over the local species will decay when the temperature is back to normal because the draining of warm water stops and the stream basin is adjusted to the water flow, which will be reinstated by the help of feeders, whose water shall be purified from noxious substances.[47]

Another protuberant example of recultivation is the surface mining lake in the former pit of Garzweiler II. Subsequent to an affluxion from the rivers Rhine and Niers of 40 years, the 180 m deep lake shall comprise an area of more than 2300 ha. For the Rhine's flow is above 2000

---

[44] Doctor Udo Rose
[45] Bund: Friends of the Earth Germany – Braunkohle im Rheinland – das Beispiel Garzweiler II (p.6)
[46] Josef Klostermann/Stefan Kronsbein – Der Raum Maas-Schwalm-Nette: Landes- und naturkundliche Beiträge
      Chapter 3: Braunkohlentagebau Garzweiler II und Naturpark Schwalm-Nette – Der klassische Konflikt
      zwischen Ökonomie und Ökologie (p.13)
[47] Doctor Udo Rose

m^3/sec, this river's gauge will at most sink by 1cm at low tide. In view of the fact that the water level will be 65 m above the groundwater, it is hoped that its reserves will recover, simultaneously offering a biotope for various aquatic organisms.[47]

Therefore, recultivation stands a chance of renaturalising Grevenbroich's water ecosystem after a long period of negative aftermath.

3.5 Concluding Examination of Recultivation

The example of Grevenbroich reveals that the challenge of recultivation can have both, positive and negative impacts, even at the same time. In order to make the former outweigh the latter, land rehabilitation should promote the creation of new and the sustainment of existing biotopes in appropriate ways in accordance with the individual circumstances whose future prospects have to be carefully weighed up against each other. Whereas absolute restoration of the land area's status would antagonise biodiversity, a certain extend of renaturalisation is necessary to protect the local species in the Erft and its meadows.

3.6 Air Pollution

But unfortunately, recultivation is no universal remedy to every soft-coal related ecological issue. Next to its doubtlessly significant impact on global warming due to CO_2-emissions of nearly 25 000 000 tons in Grevenbroich alone[48], the mining and burning of the lignite in the Rhenish area have polluted the air in many ways, instantaneously affecting Grevenbroich. In addition to 1 200 tons of particulate matter annually ejected in the Rhineland, "the power station in Neurath alone exhales more than 670 kg of neurotoxic quicksilver"[48]. Some of the most serious implications might be the power stations' nuclear radiation, which is largely unknown. [48]

Nonetheless, RWE's clean coal concept could reduce the emissions of nitrogen oxides, dust and sulphur dioxides could be reduced by an average of around 80% in Grevenbroich.[49]

---

[48] Bund: Friends of the Earth Germany – Braunkohle im Rheinland – das Beispiel Garzweiler II (pp.9/10)
[49] DEBRIV – 125 Jahre DEBRIV: Braunkohle im Zeitraum 1985 – 2010: Rohstoff/Mensch/Natur/Technik
    Part 3: Braunkohle im Spannungsfeld – Versorgungssicherheit und Klimaschutz (2005-2010-2020)
      Chapter 2: Clean Coal Concept (p.121)

4. CONCLUSION

To come to a conclusion, Grevenbroich has in every sense been shaped by lignite mining.

On the one hand, the resulting industrialisation as well as the successional industries and better working agriculture have combined to lead to a certain wealth in the municipality which was accompanied by improved infrastructure and a generally higher standard of living.

This development has also become manifest in mining companies and the allocated industries as main employers offering high wages, good working conditions and the opportunity of promotion, thereby letting Grevenbroich's population have a share of the city's newly gained importance as the energy capital of Germany.

On the other hand, the shady sides of forced relocation and the resulting loss of homes and cultural heritage could hardly be illuminated by the light from gained electricity. Comparably, Grevenbroich's ecological environment has also suffered from the unconscionable thrive for this black gold. It is doubtful whether the enriched biodiversity due to the establishment of newly structured habitats within the scope of shifting can compensate the cutback of other species. In spite of the positive renaturation, the taken procedures have all the same improved the situation in deteriorated regions and counterproductively hindered the advancement in others.

Apart from this, recultivation has brought absolutely new perspectives. With regard to the lake in Garzweiler II, it can be seen as an attempt to align economic with ecological use. Whilst the one bank shall accommodate new residential estates and recreation areas with leisure activities which might even boost tourism, the other side will be reserved to natural progression, leaving space for the region's flora and fauna.

As for RWE's role as Grevenbroich's most important entrepreneur, this can also be seen ambivalent. Anyway, the necessary deactivation of the soft coal power stations as inopportune fossil fuels in times of climate change might mean the reduction of jobs.

However, innovation in sectors like renewable energy and recultivation create new workplaces.

All in all, the influence of lignite mining on Grevenbroich has been predominantly positive so far. Still, it indeed is about time to leave the image of the capital of fossil energy behind in order to tackle future challenges broadening Grevenbroich's horizon for further gainful advance.

5. APPENDIX

i These numbers are based on the following conversion:
Around 1700, 1 Gulde used to be a master craftsman's wage for two days of 13 ½ working hours each (cf. Schön: Deutscher Münzkatalog 18. Jhdt: 1700-1806; 2008, p. 12). According to the "Ministerium für Arbeit, Gesundheit und Soziales des Landes Nordrhein-Westfalen" the basic salary of this profession amounted to hourly earnings of 22.59 € in February 2019.

$$1 \text{ Gulde} / 27 \text{ h} \triangleq 22.59 \text{ € / h}$$
$$\Leftrightarrow 1 \text{ Gulde} \triangleq 22.59 \text{ € / h x } 27 \text{ h}$$
$$\Leftrightarrow \mathbf{1 \text{ Gulde} \triangleq 609.82 \text{ €}}$$

For 1 Reichstaler equates 1 ½ Gulds, one can calculate a Reichtaler's purchasing power through this hedge formula:

$$1 \text{ Reichstaler} \triangleq 3/2 \text{ x } 609.82 \text{ €}$$
$$\Leftrightarrow \mathit{1 \text{ Reichstaler} \triangleq 914.73 \text{ €}}$$

ii Dr. Johannes Demmer on behalf of Landesplanungsgemeinschaft Rheinland - Rheinisches Braunkohlengebiet: *Wirtschafts- und Sozialstruktur Teil 1* (1962)
      Chapter 1.3(1): Die Folgeindustrien/ Allgemeine Übersicht (p.18)

iii Statistische Rundschau für den Landkreis Grevenbroich (Düsseldorf, 1968)
      Chapter 2.2 Entwicklung der Einwohnerzahlen seit 1816 (p.16)

Taten und Fakten – *Kreispolitik im Ballungsraum 1965-1969*
      Part B: Daten über den Kreis und die Kreisverwaltung
            Chapter 2: Entwicklung der Wohnbevölkerung (p.43)

Landesamt für Datenverarbeitung und Statistik Nordrhein-Westfalen; Statistische Rundschau für die Kreise Nordrhein.Westfalens; Kreis Neuss
      Part 2: Bevölkerung und Wohnverhältnisse
            Chapter 4: Natürliche Bevölkerungsbewegung und Wanderung 1975 bis 1979 (p.71)

Statistisches Jahrbuch 1989; Kreis Neuss, Der Oberkreisdirektor
      Part 2: Fläche und Bevölkerung
            Chapter 14: Einwohnerentwicklung 1980-1989 Dormagen/Grevenbroich/Meerbusch (p.48)

Statistisches Jahrbuch 1995; Kreis Neuss, Der Oberkreisdirektor
      Part 2.2: (Fläche und) Bevölkerung
            Chapter 7: Stadt Grevenbroich 1990-1995 (p.41)

Kreis Neuss; Der Landrat – Amt für Entwicklungs-, Landschaftsplanung, Wirtschaft und Statistik 2000 (2000)
      Part 1.8 Daten
            Chapter 5: Bevölkerung in den Städten und Gemeinden zum 31.12.1999 (p.29)

Rhein Kreis Neuss; Statistisches Jahrbuch 2010-Zahlen-Daten-Fakten
Part 1.15: Daten; Chapter 3: Bevölkerungsentwicklung in den Städten und Gemeinden

iv RWE Power im Rheinischen Braunkohlerevier (p.6)

6.BIBLIOGRAPHY

Printed Sources:

Dickmann, Franz (1996): Umsiedlungsatlas des rheinischen Braunkohlereviers; Bonn, Landschaftsverband Rheinland, Amt für Rheinische Landeskunde

Geologische Landesamt Nordrhein-Westphalen (1977): Tagebau Hambach und Umwelt – Auswirkungen eines geplanten Tagebaus im Rheinischen Braunkohlenrevier

Dr. Johannes Demmer on behalf of Landesplanungsgemeinschaft Rheinland - Rheinisches Braunkohlengebiet: Wirtschafts- und Sozialstruktur Teil 1 (1962)

Westdeutsche Wirtschaftsmonographie Folge 2: Braunkohle (1957)

Statistische Rundschau für den Landkreis Grevenbroich (Düsseldorf, 1968)

Taten und Fakten – Kreispolitik im Ballungsraum 1965-1969 (1969)

Landesamt für Datenverarbeitung und Statistik Nordrhein-Westfalen; Statistische Rundschau für die Kreise Nordrhein-Westphalens; Kreis Neuss (1979)

Statistisches Jahrbuch 1989; Kreis Neuss, Der Oberkreisdirektor (1989)

Statistisches Jahrbuch 1995; Kreis Neuss, Der Oberkreisdirektor (1995)

Kreis Neuss;/Der Landrat/Amt für Entwicklungs-, Landschaftsplanung, Wirtschaft und Statistik – Statistisches Jahrbuch 2000 (2000)

Rhein Kreis Neuss; Statistisches Jahrbuch 2010-Zahlen-Daten-Fakten (2010)

Dr. Peter Zenker – Braunkohle, Kraftwerke, Briketts: Der Norden des Rheinischen Braunkohlereviers (2010)

Ministerium für Ernährung, Landwirtschaft und Forsten des Landes Nordrhein-Westfalen – Wasserwirtschaftliche Probleme im Erftgebiet: Zum Gesetzesentwurf über die Gründung des großen Erftverbandes (1956)

Josef Klostermann/Stefan Kronsbein – Der Raum Maas-Schwalm-Nette: Landes- und naturkundliche Beiträge; Chapter 3: Braunkohlentagebau Garzweiler II und Naturpark Schwalm-Nette – Der klassische Konflikt zwischen Ökonomie und Ökologie (1996)

Deutscher Braunkohlen-Industrie-Verein E.V. 1885-1960 (1960)

DEBRIV – 125 Jahre DEBRIV: Braunkohle im Zeitraum 1985 – 2010: Rohstoff/Mensch/Natur/Technik (2010)

Magazines:

Braun, W.,Schneider, K.G; Weiss, G (1996): Braunkohlentagebau und Umsiedlung im Rheinischen Revier, Köln, Geostudien

RWE Power: Umsiedlung im Rheinland – Partnerschaft sichert Sozialverträglichkeit

NGZ: Tim Kronner - Restloch in Grevenbroich: Tagebau wird zur Naturlandschaft (2018)

Stattblatt: Christina Faßbender: Zukunft ohne Kohle? (2019)

Scientific Publications:

Prof. Dr. Kunz, Werner (2004): The Brown Coal Open-Cast Mining as a Place for Resettlement of Rare Butterflies and Other Organisms: What will be Destroyed by Recultivation, Düsseldorf

W. Konold, R. Böcker, U. Hampicke – Handbuch Naturschutz und Landschaftspflege
    Chapter 7: Land- und Wassernutzung und Naturschutz
    Subtopic 10.1: (Ulf Dworschak, Udo Rose-) Das Rheinische Braunkohlerevier

Bund: Friends of the Earth Germany – Braunkohle im Rheinland – das Beispiel Garzweiler II

Alexander Weuthen Die Geschichte des Braunkohletagebaus im rheinischen Braunkohlerevier und seine ökologischen und sozialen Auswirkungen (2015)

Online Sources (March, 24, 2019; 7:59 pm):

Stadt Grevenbroich: quoted from: https://www.grevenbroich.de/stadtportrait/stadtgeschichte/

Erasmus Gymnasium Grevenbroich: quoted from:
https://www.erasmus.de/historischer%20rueckblick.htm

Syntaxonomie nach POTT 1995 – Pflanzengesellschaften: quoted from:

https://gastein-im-bild.info/oeko/ob_t3.html#t34

Ministerium für Arbeit, Gesundheit und Soziales des Landes Nordrhein-Westfalen: quoted from:

http://www.tarifregister.nrw.de/tarifinformationen/grundverguetung_berufe/index.php?show=TWV
pc3Rlci8taW4=

Personal Interviews:

Dr. Udo Rose, head of the Erftverband

Prof. Werner Kunz, biology professor at the Heinrich-Heine-Universität in Düsseldorf

Dr. Gero Vinzelberg, open pit planner of RWE Power

Due to the topic's locational tie most sources are in German. Direct quotations are as literally translated into English as possible.

YOUR KNOWLEDGE HAS VALUE

- We will publish your bachelor's and
  master's thesis, essays and papers

- Your own eBook and book -
  sold worldwide in all relevant shops

- Earn money with each sale

Upload your text at www.GRIN.com
and publish for free